Snake Fights, Rabbit Fights, & More

A BOOK ABOUT ANIMAL FIGHTING

By Miriam Schlein · Illustrated by Sue Thompson

CROWN PUBLISHERS, INC., NEW YORK

10   9   8   7   6   5   4   3   2   1

The text of this book is set in 14 point Times Roman. The illustrations are black-and-white line drawings.

Library of Congress Cataloging in Publication Data
Schlein, Miriam.   Snake fights, rabbit fights, and more.   Bibliography: p. 48
Summary: Explores the reasons, including food, nesting places, mates, rank, and territory, why animals display aggressive behavior and fight with other animals like themselves. 1. Animal fighting—Juvenile literature. 2. Aggressive behavior in animals—Juvenile literature. [1. Animal fighting. 2. Aggressive behavior in animals. 3. Animals—Habits and behavior] I. Thompson, Sue. II. Title.   QL758.5.S3   1979   596'.05   79-2340   ISBN 0-517-53417-7

Snake Fights,
Rabbit Fights, & More

What's happening here?

What are these rabbits doing?

They are facing each other and circling around. One gives the other a smack with his paw. Then they leap into the air and throw their bodies against each other. They bite and kick each other with their big strong hind feet. They are having a terrific fight.

What started it?

It is the breeding season. At this time, one male rabbit will mate with several female rabbits. And he will fight fiercely to chase away any other male who comes close and wants to mate with these females.

A female rabbit will fight, too, if another female comes too near her nest and babies.

Wherever animals live in the wild, there are all kinds of animal fights. We don't mean one kind of animal killing a different kind of animal so he can eat—like a fox killing a rabbit. We mean an animal fighting with an animal like itself.

Why do animals fight?

Fighting for a mate—as the rabbits were doing—is the reason for many animal fights. But there can be other reasons. Animals fight over food. They fight over nesting places. If they live in groups, they fight over rank, or their position in the group. Or they fight over territory.

What are these fish doing?

Are they kissing each other?

No. They're fighting, too.

These little fish are called cichlids. (This sounds like sick-lids.)

They are fighting because one came swimming into the other's territory.

One fish lives here. This small part of the lake has everything he needs. It has the kind of food that he eats. And it has a sandy bottom where he can make a hollow nest by brushing the sand with his tail. His mate will lay her eggs in this nest.

In a way, he considers this part of the lake his very own. Scientists call this part of the lake the fish's *territory*.

Other types of fish can live there, too. But if the cichlid sees another cichlid like himself swimming into his territory, he will fight to keep him out—not just in breeding season but any time.

He doesn't start to fight right away. First he just tries to scare the other fish away by making himself look bigger. He spreads

out his fins. Scientists call this *display*. Now the other fish can see the bright-colored markings on the inside of his fins. This shows he is ready to fight.

Then, with his tail, he swishes water at the other fish. If the other fish still doesn't go away, then they begin to fight. They grab each other by the mouth and have a tug-of-war. They pull each other back and forth. It doesn't hurt because they have thick, tough, leathery skin around their mouths.

Finally one of the fish lets go and slowly swims away. He has given up.

Most often, the fish who is defending his own territory is the winner.

*B*irds fight over territory, too.

This robin has just been in a fight. That is why he looks ruffled and tired. He has just chased away another robin who wanted to come into his territory.

His territory is not very big. But it has everything he needs. It has berry bushes and trees and insects. That is why he picked it out. Now, in the center of the territory, he and a female robin have a nest and baby robins.

If he did not fight to keep other robins away, too many robins might settle too close to each other. There would not be enough food nearby for all of them. This would mean the father and mother would have to fly a greater distance from the nest each time they went to find food.

They take turns guarding the nest. The mother as well as the father will fight to keep strangers away from the nest. And it is better if both are close to the nest most of the time.

With a territory that they do not share with other robins, it is easier for the mother and father robin to feed and protect their young. And there is a much better chance that their young will grow up safely.

Different kinds of birds need different-size territories. Robins need a territory only about 100 by 100 feet. (About 33 by 33 meters.) Larger birds—owls or eagles—who hunt for prey, such as mice and rabbits, need much larger territories in which to hunt.

A horned owl's territory is about 2,300 by 2,300 feet. (About 700 by 700 meters.)

A bald eagle needs a territory even larger than that.

But an Adélie penguin in Antarctica, sitting on its nest of stones, protects a territory only as far as its beak can reach.

How can an animal tell that he might be entering a territory that is already taken? It's easy. Birds and animals leave signs. You may not understand these signs, but the birds and animals do.

Some animals leave signs that can be smelled. Many kinds of deer and antelope have scent glands near their eyes. They leave secretions from these glands on branches.

Dogs and wolves urinate around the border of their territory. That is their way of marking off their territory.

Some animals leave signs you can see. A hippopotamus leaves a sign that nobody can miss. It's a big pile of hippo turds. It can be 5 feet wide and 3 feet high. He leaves it on his trail, which leads from the river to the place on land where he gets his food. It is his sign that tells other male hippos to stay away.

Others have signs you can hear.

The robin sings. His song is a warning to keep other robins away. It is as though he is saying: I am here. I am a robin, and this place is mine.

Other robins understand this. But if another robin hears the robin sing and comes in anyway, there will probably be a fight.

Howling monkeys do something else. They howl. This is their way of telling other howlers that this part of the jungle is theirs. They howl so loud that another group of howlers can hear them miles away.

*H*ave you ever seen a snake fight?

It is not very violent. In fact, it looks more like a dance than a fight. Scientists have given it a special name. They call it the Male Combat Dance.

Scientists do not know exactly why snakes do the Male Combat Dance. Most snake experts believe it has something to do with sex and mating. Perhaps one male snake is trying to keep another male away from the female he is courting.

Here are two rattlesnakes fighting. This is what they do. They don't use their fangs at all. They have a special way of fighting. They twist the bottom part of their bodies around each other. They lift their heads and hold them side by side. Then they push very hard. Each snake tries to push the other snake's head sideways down to the ground and keep it pinned there for a short time. The one who does this is the winner. Then they disentangle. The loser glides off.

Other times, the fight is slightly different. They face each other, raise the top third of their bodies, and rub their underscales against each other. They press till one loses balance. Or, sometimes, one snake quickly twines himself two times around the other snake's upper body and throws him down to the ground. The snake who does this is the winner. The loser glides off.

Did you wonder why they did not use their fangs in the fight?

They did not use their fangs because they were obeying an important animal "fighting rule." They did not want to kill each other. This is true of most animal fights. They do not fight to the death.

Instead, many animals have a very special way of fighting. There are no sneak attacks or surprises. The fight follows a certain pattern. They do certain things and always in the same order. It is as though the fighters are playing a game and following definite rules.

Scientists have a name for this. They call it ritualized fighting. The word *ritual* means a ceremony that is always performed in the same way, with certain steps happening in a certain order.

These ritualized fights take place instead of fierce anything-goes kinds of fights, in which the animals use all their weapons —the kind of fight that would end only when one or both of the animals was painfully wounded or killed.

These rituals have evolved in the behavior of animals over millions of years. They are nature's way of preventing too many members of a species from killing one another off.

*L*ook at these two fallow deer. They are males, and it is the mating season. What they are doing looks very odd, if you don't know that they are having a ritualized fight.

They are walking side by side, lifting their legs very high. It looks as though they are in a parade.

Their heads face straight ahead. But each deer is looking at the other out of the corner of his eye. They are nodding their heads up and down slowly, showing off their antlers to each other. Their antlers are their fighting weapons.

When the parade is over, they turn and face each other. They bend their heads and push with their antlers. They push to see who is the stronger one. If neither one seems to be stronger, they stop and parade again. Then they have a second round of pushing.

Finally, one shows that he can push longer and harder than the other. He is the winner. He will be the one to mate with the group of female deer. The other male leaves.

Sometimes, in the parade, their timing is off. One deer will turn, all ready to push. But the other deer is still high-stepping along and waving his antlers.

The first deer may have his head lowered and his antlers pointing at the unprotected side of the other deer. But he will not attack. Instead, he will take a few fast steps, catch up to the other deer, and join him again in the parade. When they are both ready, they will hold the next round.

They have these fights in the mating season. At other times of the year, deer wander together in herds and graze without fighting.

Sometimes deer do die because of a fight. But it is an accident. Their antlers may become entangled, and if they can't get them apart, they won't be able to run or protect themselves from an enemy. So they might be attacked, or they might die of starvation.

Sometimes the skeletons of two deer are found with their antlers still entangled.

This big lizard on top of the rock is guarding his territory. He is a four-foot-long iguana. At the base of the rock is another iguana guarding *his* territory.

Most of the year, all the iguanas live together on the beach. But when the breeding season comes, the males stake out separate territories. Space is limited along the shore, so their territories are small.

Several females stay with a male in his territory, and he mates with them. If another male steps into his territory, he will fight to chase him away. But he doesn't start fighting right away.

First he tries to scare the other one away. He makes himself look bigger. The crest along his neck and back stands up. And he stands higher on his legs.

Then he opens his mouth and moves slowly toward the intruder, nodding his head up and down. If the intruder still doesn't leave, they fight. They lower their heads and push. The horny ridges on top of their heads mesh with each other. This keeps their heads from slipping during the fight. They push until one iguana pushes the other one out of the territory.

Female iguanas fight each other, too. They lay their eggs in the sand, and if good nesting places are scarce, they fight over the best ones.

When male iguanas fight, the one who is defending his own territory is usually the winner. But this does not happen all the time. Sometimes the intruder is bigger or younger and stronger. If he can push the first iguana out, then he stays and takes over the territory and mates with the females. The first iguana leaves.

But that did not happen here. Sometimes the intruder can tell right away that he is going to be pushed out. So instead of letting this happen, he gives up.

How does he let the other iguana know he is giving up?

He does it with his body. He lies down flat, stretches out his legs, and lowers his head. All this makes him look small and weak.

The other iguana knows what this means. He stops fighting and just waits until the intruder creeps away.

Different animals have different ways
of showing that they give up.
A rooster will stick his head in a pail or a hole.

A gull hides his head in his feathers.
Many animals will lie flat on the ground
with their legs stretched out and their heads
down, the way the iguana does.
A gnu will do this.
So will a rabbit.

If a wolf sees he is losing a fight and wants to give up, he bows his head, turns it to one side, and exposes his neck—the worst place he could be bitten—to his opponent. This is his way of saying: I am in your power. You win.

And that will stop the fight.

A dog does the same thing.

Sometimes we can see these fighting rules in action.

THE DOG FIGHT

Not long ago, there were two dogs that got into a terrible fight. Max was bigger than Fog. But Fog was an older and more experienced dog.

They growled and snarled and rolled over and over each other, snapping and biting. Nobody could get near them. Then Max's owner saw that Fog had his teeth around Max's neck.

"He's going to kill him!" she screamed.

But suddenly Max bowed his head and let it hang limp. Slowly,
Fog withdrew his teeth from around Max's neck.

The fight was over.

Max had used the animal signal to say *I give up*.

And Fog obeyed the signal.

The fighting rule had saved Max's life.

When a hippo gives a big yawn—watch out! That is, if you're another hippo. It doesn't mean that he is sleepy or bored. It means that he is ready to fight. What he is doing is showing his teeth.

Females and young hippos live together in a group in the river. Scientists call this the *creche*. The males stay in territories around the creche. If one male enters another's territory, trying to make his way to the females, a fight begins.

At first they face each other, meeting mouth to mouth. This is the threatening part of the fight. Each one wants to see if the other really intends to fight. If the fight does go further than this, it can become bloody and violent. The hippos stand side by side in the shallow water, facing opposite ways. They open their mouths and swing their heads at each other, trying to gash each other in the side with their tremendous curved teeth.

The hippo is one of the few animals that will often seriously hurt or even kill his opponent. Often old hippos are seen with big scars all over their bodies. Sometimes a piece of tooth breaks off and becomes embedded forever in the hippo's hide.

If a hippo's front leg is broken in a fight, he will not be able to get up on land to get his food and he will starve to death.

Or if a tooth pierces his heart during a fight, he will die. But most fights end when one hippo decides he has had enough. Then he throws himself into the water with screams of pain and rage.

Some of the weaker males don't want to go through all this. So they just go off to live alone in a big puddle or an inland pool.

Sometimes rats kill one another in a fight, too.

Wild Norway rats live together in a large pack. They usually get along with one another. But they will fight a stranger if he tries to join their pack. They can tell he is a stranger by his smell. And one of the rats in the pack will fight to keep him out.

That is what started this fight. Look at them. The two rats are up on their hind legs, and they are punching with their forelegs. It's like a boxing match. Now they're kicking, too.

It's a noisy fight. They gnash their teeth and cry out. Finally, the rat from the pack gives a very hard punch. The stranger tumbles onto his back. Then he scrambles up and runs away.

Rats usually do not bite during the fight. But sometimes one rat will get desperate and begin to bite. Then the other rat begins to bite, too. When this happens, the fight often ends in death for one or both of the rats.

Some animals who live in groups fight for another reason. They fight over *rank*, or position in the group.

A male giraffe will fight over rank with other males in the group. He uses his head as a weapon.

Here a fight is just beginning. Two males are standing shoulder to shoulder.

They push and circle around. Then one stands back and swings his head against the other giraffe's neck or side.

The second giraffe braces himself. Sometimes the blow is so strong that his forelegs are lifted off the ground.

Now it's his turn. He swings his head at the other giraffe.

When one giraffe has had enough, he gallops away. The winner runs after him, holding his head high.

The loser comes back in a few minutes. Now, whenever he passes the other giraffe, he lowers his head a bit—like a salute.

A giraffe can give a powerful kick with his hoof, but he will never use his hoofs or horns against another giraffe. He will use them only against an enemy, like a lion who is attacking him.

*R*ank is important for many animals who live in groups. Animals of higher rank usually lead the group. They also mate more with the females.

A high-ranking baboon will break up small fights between other members in the troop. He will also step up and protect a member of his troop who is threatened from the outside.

Not every single fight between two animals is a ritualized fight. Male lions will fight and sometimes kill each other. Hyenas will fight over a carcass they are eating. If one hyena kills the other, he will often eat him as well.

When animals do not have enough space, more fighting goes on.

Spider monkeys shake branches at each other when they are angry. They also kick, slap, and bite.

A troop of baboons will peacefully fan out over the ground, looking for grass shoots to eat. But if the food is in one clump, they will fight over it.

Sometimes a fight that begins as a ritualized fight will turn into a violent fight. Then one or both animals may be seriously hurt or even killed. A fish, for example, might ram into the side of the other fish.

Fighting plays an important and necessary part in animals' lives. In most cases, the animals are not badly hurt, if at all. And in most cases, the results of the fighting are good.

Territorial fights help to keep too many of the same species from living too close to one another, all competing for the same food and space.

Fighting to establish rank helps the animals to form some sort of social organization. Having rank in the group and having a leader helps the group to avoid many small arguments. It also helps the group to present a united front when threatened by outside enemies.

And the fights that occur when males are competing for a mate have results that reach far into the future.

The stronger animal wins. So it is the strongest males who get to mate and be the fathers of the next generation.

This is good in two ways. A stronger father is better able to protect his young, if they belong to a species in which the father helps to care for the young.

Also, a strong father is more likely to have strong and healthy young.

When two females fight for the best nesting spots, the stronger

one also wins. A strong mother is more likely to give birth to strong, healthy young. And a strong mother will also be better able to care for her young so that they have a better chance to grow to adulthood.

As this keeps going on, generation after generation, over hundreds, thousands, and sometimes millions of years, the entire species becomes stronger.

And so, as all these fights go on in the animal world—in fields, in jungles, or under the water—most often they do not lead to death or serious wounds. Instead, most often, these fights lead to life and a new generation.

BIBLIOGRAPHY

DORST, JEAN. *The Life of Birds*. New York: Columbia University Press, 1974.

DROSCHER, VITUS B. *The Mysterious Senses of Animals*. Translated from the German by Eveleen Huggard. New York: E. P. Dutton and Company, Inc., 1965.

EIBL-EIBESFELDT, IRENÄUS. *The Fighting Behavior of Animals*. *Scientific American* (December 1961), pp. 112-122. CCV.

———. *Love and Hate*. New York, Chicago, San Francisco: Holt, Rinehart and Winston, 1971.

Encyclopedia of Reptiles, Amphibians, and Other Cold-Blooded Animals. Introduction by Dr. Maurice Burton. "Octopus," in association with Phoebus. Distributed in the United States by Crescent Books, a Division of Crown Publishers, Inc., One Park Avenue, New York, N.Y. 10016. London: BPC Publishing, Ltd., 1968/69/70. London: Octopus Books, Ltd., 1975.

Grzimek's Animal Life Encyclopedia. Edited by Dr. Bernhard Grzimek. New York: Van Nostrand Reinhold Co., 1972/73/74/75.

Grzimek's Encyclopedia of Ethology. Edited by Dr. Bernhard Grzimek. New York, Toronto, London, Melbourne: Van Nostrand Reinhold Co., 1972/73/74/75.

KLAUBER, LAURENCE. *Rattlesnakes*. Berkeley, Calif.: Published for the Zoological Society of San Diego by the University of California Press, 1972.

LORENZ, KONRAD. *On Aggression*. Translated by Marjorie Kerr Wilson. New York: Harcourt, Brace and World, Inc., 1966.

WELTY, JOEL CARL. *The Life of Birds*. Philadelphia, London, Toronto: W. B. Saunders Co., 1975.

WHEELER, ALWYNE. *Fishes of the World*. New York: Macmillan Publishing Co., Inc., 1975.

WILSON, EDWARD O. *Sociobiology, the New Synthesis*. Cambridge, Mass., and London, England: The Belknap Press of Harvard University Press, 1975.